Cambridge Elements ⁼

Elements in Geochemical Tracers in Earth System Science
edited by
Timothy Lyons
University of California
Alexandra Turchyn
University of Cambridge
Chris Reinhard
Georgia Institute of Technology

VANADIUM ISOTOPES

A Proxy for Ocean Oxygen Variations

Sune G. Nielsen

Woods Hole Oceanographic Institution, Massachusetts

CAMBRIDGE
UNIVERSITY PRESS

CAMBRIDGE
UNIVERSITY PRESS

University Printing House, Cambridge CB2 8BS, United Kingdom

One Liberty Plaza, 20th Floor, New York, NY 10006, USA

477 Williamstown Road, Port Melbourne, VIC 3207, Australia

314–321, 3rd Floor, Plot 3, Splendor Forum, Jasola District Centre,
New Delhi – 110025, India

79 Anson Road, #06–04/06, Singapore 079906

Cambridge University Press is part of the University of Cambridge.

It furthers the University's mission by disseminating knowledge in the pursuit of
education, learning, and research at the highest international levels of excellence.

www.cambridge.org
Information on this title: www.cambridge.org/9781108797948
DOI: 10.1017/9781108863438

First published 2020

A catalogue record for this publication is available from the British Library.

ISBN 978-1-108-79794-8 Paperback
ISSN 2515-7027 (online)
ISSN 2515-6454 (print)

Vanadium Isotopes

A Proxy for Ocean Oxygen Variations

Elements in Geochemical Tracers in Earth System Science

DOI: 10.1017/9781108863438
First published online: December 2020

Sune G. Nielsen
Woods Hole Oceanographic Institution, Massachusetts
Author for correspondence: Sune G. Nielsen, snielsen@whoi.edu

Abstract: Vanadium isotope ratios ($^{51}V/^{50}V$) have potential to provide information about changes in past ocean oxygen contents. In particular, V isotopes may find utility in tracing variations at nonzero oxygen concentrations because the redox couple that controls V elemental and isotopic abundances in seawater (vanadate–vanadyl) appears to operate around 10 μM O_2. This characteristic sets V isotopes apart from many other metal isotope redox proxies that require more reducing conditions to register significant changes in their isotope budgets. The oxygen abundance sensitivity range of V isotopes suggests that this paleo-proxy could be particularly useful in tracing marine oxygenation changes throughout the Phanerozoic and potentially beyond.

Keywords: Vanadium isotopes, ocean oxygenation, redox sensitive, marine mass balance

ISBNs: 9781108797948 (PB), 9781108863438 (OC)
ISSNs: 2515-7027 (online), 2515-6454 (print)

Contents

1 Introduction

There is great interest in understanding how the concentration of oxygen evolved in the ocean and atmosphere over geological time because it has direct links to, for example, the evolution of photosynthesis, emergence of macroscopic life forms, and mass extinction events. Most commonly, evolution of the marine oxygen reservoir is studied via chemical or biomarker proxies that are sensitive to changes in the redox state of the sediment depositional environment (Canfield et al., 2018; Tribovillard et al., 2006). It has long been recognized that the concentrations of different trace metals in sediments depend strongly on the redox state of the overlying water column as well as the overall supply rate of the metal to seawater (Tribovillard et al., 2006). In turn, marine input fluxes can depend strongly on atmospheric oxygen concentrations, which can be independent of water-column oxygen concentrations (Anbar et al., 2007). However, trace metal concentrations in low-oxygen sediments are also controlled by the mechanisms of enrichment, which are affected by local processes such as basinal restriction and redistribution of Fe and Mn oxides from shallower, oxic portions of the seafloor (Algeo, 2004; McArthur et al., 2008; Ostrander et al., 2019b; Owens et al., 2012). Hence, trace metal concentrations alone can in principle be interpreted in terms of changes in either the oceanic or the atmospheric oxygen reservoirs or, alternatively, changes in the local depositional environment, depending on which process(es) each element is controlled by.

In the last two decades, stable isotopic studies of a host of these trace metals, like Fe, Mo, U, Tl, Se, and Cr (Frei et al., 2009; Kendall et al., 2017; Lau et al., 2019; Ostrander et al., 2019a; Stüeken et al., 2015), have been utilized to gain further insight into past oxygen fluctuations because stable isotope variations, in concert with concentrations, potentially can be used to disentangle effects from ocean–atmosphere redox and local depositional effects, which is more challenging with concentrations alone. Metal stable isotope fractionation is typically most pronounced for isotopic equilibria between species with different redox states (Bigeleisen and Mayer, 1947), which is the reason why most of these isotope systems have found utility as redox proxies. Such species equilibria, or redox couples, occur at different levels of redox potential (Rue et al., 1997), such that metal stable isotope variations in sediments are likely indicators of specific levels on a redox ladder (figure 1 in Owens, 2019). Essentially all the metals utilized to date as paleo-redox proxies have redox couples significantly below the O_2–H_2O redox couple. Therefore, these metal isotope proxies may provide information about past changes in ocean redox but can be difficult to utilize as proxies for how much oxygen was present at the time of deposition.

However, many geologically important events, such as ocean anoxic events (OAEs), the Permo-Triassic mass extinction, and the Cambrian explosion, all occurred at times when the oceans were likely fully or at least partially oxygenated relative to today (Jenkyns, 2010; Lu et al., 2018; Lyons et al., 2014). Thus, changes in ocean oxygenation during these periods likely transitioned across moderately oxygenated conditions, which may be difficult to detect with existing isotope proxies.

The element vanadium (V) potentially is different from other metals used as isotope redox proxies because it consists of multiple redox couples, the most oxic of which may operate at nonzero oxygen concentrations (Gustafsson, 2019; Wehrli and Stumm, 1989). Specifically, V can occur as V(III), V(IV), and V(V) in different marine sediment depositional environments (Gustafsson, 2019; Wehrli and Stumm, 1989). In open ocean, fully oxygenated seawater, V is exclusively present as vanadate (VO_4^{3-}) or its hydrolyzed versions (HVO_4^{2-}, $H_2VO_4^-$). However, at low oxygen, vanadate is reduced to vanadyl (VO^{2+}) or its hydrolyzed versions ($VO(OH)^+$, $VO(OH)_2$). In addition, the stability of vanadyl at higher redox potentials is augmented by the presence of organic ligands (Wehrli and Stumm, 1989). Therefore, it is likely that the VO_4^{3-}–VO^{2+} redox couple operates at nonzero oxygen concentrations, which opens the potential for using V isotopes as an indicator for the level of marine oxygenation rather than, for example, areal extent of anoxia or euxinia, as has been suggested for other metal isotope proxies.

High-precision measurements of V isotope ratios ($\delta^{51}V = 1000 \times [(^{51}V/^{50}V_{sample} - {}^{51}V/^{50}V_{AA})/^{51}V/^{50}V_{AA}]$) have only recently been developed (Nielsen et al., 2011; Nielsen et al., 2016; Prytulak et al., 2011; Wu et al., 2016), primarily because V has only two isotopes (^{50}V and ^{51}V) and the low abundance of ^{50}V (~0.24 percent) makes it very difficult to obtain precise and accurate isotope ratios. Therefore, there is still much to learn about the conditions under which V isotopes might be used as a paleo-redox proxy, as well as which sediment archives are most appropriate for exploring V isotope variations. In the following, I first review the marine elemental and isotopic cycles for V, followed by an outline of the analytical methods required for precise and accurate V isotope measurements. Lastly, I discuss during which geological time periods and in which sedimentary archives V isotopes are most likely to find utility in the future.

2 Marine Elemental and Isotopic Cycle of Vanadium

The marine mass balance of V has been the subject of multiple previous studies (Emerson and Huested, 1991; Morford and Emerson, 1999). However, since the most recent assessment, several findings that affect the input and output fluxes of V have been published, which I here use to reassess the marine elemental V budget. Subsequently, I couple the elemental mass balance with recent V isotope data for marine sediments to construct a marine isotopic budget for V.

2.1 Elemental V Mass Balance

There are five principal marine V fluxes: rivers, hydrothermal sediments, oxic sediments, euxinic sediments, and anoxic sediments (Figure 1). The V concentration of open ocean oxygenated seawater is, based on the most recent GEOTRACES data (Ho et al., 2018), relatively invariant globally at ~35 nmol/kg, which is consistent with the first V determinations of seawater (Collier, 1984; Jeandel et al., 1987) and implies that all the major ocean basins have essentially invariant V concentrations. Surface seawater is only mildly depleted in V (~10 percent relative to deep water) and, because the minor V uptake into particulate material at the surface is remineralized at depth in oxygenated sea-water (Collier, 1984), particulates from oxic portions of the water column are unlikely to have a profound effect on the overall marine V mass balance.

Rivers constitute a major input of V to the ocean. The riverine V input flux has been investigated relatively extensively and the most recent compilation of river concentration data concluded that the global V flux to the ocean is 520 Mmol/yr (Shiller and Mao, 2000). These data also include considerations of different weathering styles, source rock compositional variations, and estuarine processes (Shiller and Boyle, 1987; Shiller and Mao, 2000), making this a relatively robust number.

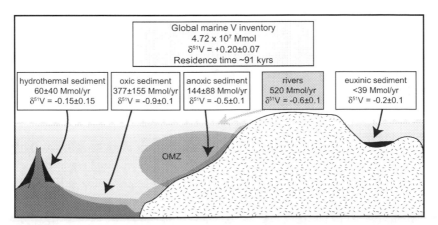

Figure 1 Summary of the fluxes and isotope compositions of the major marine inputs and outputs for V. Details of fluxes and their calculated magnitudes are described in the text. Isotope compositions taken from Schuth et al. (2019); Wu et al. (2017); Wu et al. (2019a); Wu et al. (2020).

Hydrothermal sediments represent a significant marine V sink due to sorption onto or coprecipitation with Fe oxides (Trefry and Metz, 1989). Previous estimates of this V flux reported values of 140 and 430 Mmol/yr (Rudnicki and Elderfield, 1993; Trefry and Metz, 1989). These estimates were based on the relatively invariant V/Fe $\sim 4 \times 10^{-3}$ in hydrothermal sediments (Trefry and Metz, 1989). When this ratio was combined with the average Fe concentration of endmember hydrothermal fluids ($\sim$1 mmol/kg), an average Fe precipitation efficiency of 50 and 100 percent, respectively, and a hydrothermal fluid flux (HT$_{FluidFlux}$) of 7 and 10×10^{13} kg/yr, respectively, then the two estimates were obtained. The best estimates for the average V/Fe ratio of hydrothermal sediments (V/Fe$_{HT-seds}$) and Fe concentrations in hydrothermal fluids (Fe$_{HT-fluid}$) remain the same as previously, and the best estimate for Fe precipitation efficiency (Fe$_{PrecEff}$) is likely still $\sim$100 percent (Rudnicki and Elderfield, 1993), even considering that significant amounts of hydrothermal dissolved Fe can be traced over large distances in the ocean (Fitzsimmons et al., 2014; Fitzsimmons et al., 2017; Tagliabue et al., 2010). However, the most recent estimates for high-temperature hydrothermal fluid fluxes are substantially lower than earlier values. The most comprehensive recent estimates reveal values of $\sim$0.6–2.5 $\times 10^{13}$ kg/yr (Coogan and Dosso, 2012; Nielsen et al., 2017; Teagle et al., 2003), about an order of magnitude lower than what was previously used to calculate V fluxes into hydrothermal sediments (V$_{HT-flux}$). As in previous studies, the parameters mentioned are combined in the equation:

$$V_{HT-flux} = (V/Fe)_{HT-seds} \times Fe_{HT-fluid} \times Fe_{PrecEff} \times HT_{FluidFlux} \qquad (1)$$

This yields a best estimate of 60 ± 40 Mmol/yr for the hydrothermal V output flux if we implement a hydrothermal fluid flux of $1.5 \pm 1 \times 10^{13}$ kg/yr, which renders hydrothermal V output fluxes significantly smaller than previously thought.

Oxic sediments are known to accumulate substantial amounts of V through adsorption onto Fe oxyhydroxides and Mn oxides and constitute an output flux. An earlier estimate separated oxic sediments into pelagic clays/carbonates and ferromanganese (Fe-Mn) crusts/nodules and listed fluxes of 430 and 80 Mmol/yr (Morford and Emerson, 1999), respectively, although the method for calculating these fluxes was not detailed. Here, published authigenic V fluxes to pelagic sediments of 0.12 ± 0.05 µmol/cm^2 $\times$ kyr (Thomson et al., 1984) are used in combination with the total area of pelagic sedimentation of 3.11×10^{18} cm^2 (Southam and Hay, 1977). This yields a V flux associated with pelagic clays of 373 ± 155 Mmol/yr, which is within error of the previous estimate for pelagic clays. However, further studies of the V flux into pelagic clays would be likely to constrain this number significantly better. Ferromanganese crusts and nodules have V concentrations of ~400–800 µg/g (Hein et al., 2003; Wu et al., 2019b), and the total mass of Fe-Mn crusts and nodules on the ocean floor has been estimated to be $\sim 1 \times 10^{18}$ g (Hein et al., 2003). Using a very conservatively low average age for the base layer of these deposits of 2 Ma (Rehkämper and Nielsen, 2004) results in an annual V flux associated with Fe-Mn crusts and nodules of ~6 ± 2 Mmol/yr. This value should be seen as an absolute maximum, given that most of these Fe-Mn deposits are likely older than the 2 Ma used here. Hence, V fluxes associated with Fe-Mn deposits (nonhydrothermal) can be considered negligible in the global marine V mass balance.

Sediments deposited in low oxygen environments, in particular those with high organic carbon contents, are generally enriched in V relative to crustal values (Calvert and Pedersen, 1993; Morford and Emerson, 1999). Typically, these enrichments have been ascribed to diffusional transport of V across the sediment–water interface in low-oxygen conditions (Calvert and Pedersen, 1993; Francois, 1988; Morford and Emerson, 1999). However, several recent studies have investigated the V abundances in marine particles from oxic (South Pacific), anoxic (Peru Margin oxygen minimum zone or OMZ), and euxinic (Cariaco Basin) settings (Calvert et al., 2015; Ho et al., 2018; Ohnemus et al., 2017). These data reveal that marine particulates collected from oxic seawater are generally not significantly enriched relative to average crustal abundances, in agreement with earlier work (Collier, 1984). This finding was independent of organic matter, Fe, and P contents (Ho et al., 2018; Ohnemus et al., 2017), implying that biological utilization of V by primary producers is relatively minor. On the other hand, all particle samples collected from anoxic or euxinic water masses uniformly reveal substantial V enrichments relative to crustal backgrounds. Furthermore, the level of enrichment in these particles is similar to those found in organic-rich anoxic and euxinic sediments (Calvert et al., 2015). Although there is significant uncertainty in the exact process responsible for the V enrichment, most likely, vanadate is reduced at low oxygen to vanadyl species, which are subsequently incorporated into organic matter through either a biotic or an abiotic pathway (Ohnemus et al., 2017). Vanadyl species are particularly susceptible to incorporation into tetrapyrroles, which are decay products of chlorophyll; this also accounts for the high V concentrations in many crude oils (Lewan and Maynard, 1982). It is notable that several earlier studies investigating V contents in marine particles collected with sediment traps in both anoxic and euxinic waters found no enrichment in particulate V relative to crustal values (Francois, 1988; Nameroff et al., 2002). The more recent studies include particles collected using both in-situ pumps (Ho et al., 2018; Ohnemus et al., 2017) and sediment traps (Calvert et al., 2015); hence, the origin of these apparent discrepancies is not clear. However, it has previously been argued that sediment trap material can lose significant amounts of authigenic constituents before sample recovery (Kumar et al., 1996), potentially accounting for the lower V in sediment trap particles.

As outlined earlier, the stability of V(IV) species is not only dependent on oxygen concentrations; they are also preferentially stabilized in the presence of organic ligands (Beck et al., 2008; Brumsack and Gieskes, 1983; Emerson and Huested, 1991; Wehrli and Stumm, 1989), which can make it difficult to predict the oxygen content at which V is most efficiently incorporated into marine particles. However, most studies have found that anoxic (here defined as predominantly without oxygen at and above the sediment–water interface, such as the Santa Barbara Basin and the most pronounced OMZs) and euxinic (here defined as predominantly sulfidic at and above the sediment–water interface, for example observed in the Black Sea, Saanich Inlet, and Cariaco Basin) sediments are more enriched in V than suboxic sediments (here defined as sediments where oxygen penetrates <3 cm into the sediment due to lowered bottom water (BW) oxygen contents or high rates of oxygen consumption within the sediment, observed, for example, in many OMZs and coastal regions that are not strongly stratified). This general pattern of V enrichment implies that reduction of vanadate is most efficient at close to zero oxygen. In addition, it has been shown that further reduction of vanadyl to insoluble trivalent V species in euxinic environments is not kinetically favorable (Wanty and Goldhaber, 1992), which may explain why particles and sediments in euxinic basins exhibit similar enrichments to those in anoxic environments. Vanadyl may also adsorb to oxides within the sediment (Wehrli and Stumm, 1989), although the Fe and Mn oxides typically responsible for V enrichment in oxic sediments are often reductively dissolved in low-oxygen sediments, thus eliminating this process as a likely vector of V enrichment except potentially under the operation of an Fe oxide shuttle (Scholz et al., 2011).

In terms of fluxes, previous studies have inferred that V is added to the ocean via release of V-rich porewaters from suboxic sediments (Morford and Emerson, 1999; Morford et al., 2005; Shaw et al., 1990). One study concluded that 300–800 Mmol/yr of V was added to the ocean via this mechanism (Morford and Emerson, 1999). This range of values was determined by comparing V/Al ratios in bulk suboxic sediments with those found in post-Archean Australian shales (PAAS; Nance and Taylor, 1976) and observing that PAAS had higher V/Al ratios than some suboxic sediments. However, many shales are enriched in V relative to crustal values because they were deposited in low-oxygen environments, and, therefore, it has been argued (Nameroff et al., 2002) that a more appropriate comparison to assess whether suboxic sediments are enriched or depleted in V is average upper continental crust (Rudnick and Gao, 2003) or crustal compositions from the most proximal source relative to the marine sediment studied (Calvert et al., 2015; Morford and Emerson, 1999; Scholz et al., 2011). Such a comparison has previously been carried out for the Mexican Margin (Nameroff et al., 2004), where it was found that V was at crustal levels for sediments with BW oxygen concentration >10 μM. When applying the crustal value as baseline for sediments with similar BW O_2 contents from the Northeastern Pacific, the North African Margin, the Arabian Sea (Morford and Emerson, 1999), and the California Margin (Shaw et al., 1990), all samples are within error of crustal background levels (Figure 2). Based on these considerations, it must be concluded that suboxic sediments do not represent a significant marine input or output flux for V. This conclusion does not negate the fact that sediment porewaters rich in V can be released from suboxic sediments (Emerson and Huested, 1991; Morford and Emerson, 1999; Morford et al., 2005; Shaw et al., 1990). However, these fluxes of V were likely initially removed from the water column via either adsorption to oxides or diffusive sequestration by sedimentary organic matter. Hence, the net marine flux for sediments deposited under an oxygenated water column with <3 cm oxygen penetration is likely negligible.

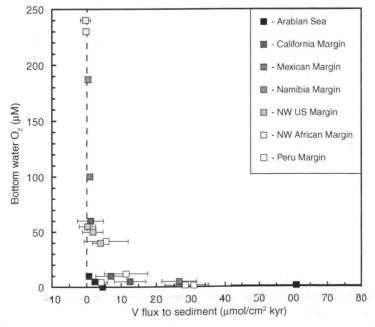

Figure 2 Fluxes of V into organic-rich sediments deposited on continental margins plotted against bottom water oxygen concentrations. Data are tabulated in Table 1.

In contrast to oxic portions of continental margins, several studies have shown that sediments deposited with BW oxygen concentrations <10 µM are significantly enriched relative to crustal backgrounds (Table 1). Here, sediment core top data are compiled from 10 sites in the Arabian Sea, the Mexican Margin, the Peru Margin, and the California Margin, and we calculate authigenic V fluxes based on V/Al enrichments relative to crustal values (Table 1). Although the range of calculated fluxes is large (2.4–60.9 µmol/cm^2 × kyr), it is notable that every site exhibits a positive V flux (Figure 2) with an average value of 18.8 ± 11.5 µmol/cm^2 × kyr (2se). When this average flux is combined with the global area of open ocean sediment deposition where the sediment–water interface has <9 µM O$_2$ (764,000 km^2, or 0.21 percent of total seafloor (Helly and Levin, 2004)), the V flux into anoxic sediments is 144 ± 88 Mmol/yr. This estimate is larger but still within error of previous estimates of anoxic V mass fluxes (Emerson and Huested, 1991; Morford and Emerson, 1999). However, in this estimate a specific limit for the amount of oxygen present at the sediment–water interface is used, which is based on published core top V fluxes. Thus, it should be more straightforward in the future to revise this estimate if additional V flux data become available.

Euxinic sediment fluxes are more complex to determine for several reasons. First, some euxinic basins (e.g. the Black Sea) are highly restricted, which may result in lower depositional rates than euxinic basins that are better connected to the open ocean (Lyons et al., 2009). In addition, a significant fraction of the V in highly restricted basins likely originates from local continental runoff rather than from open ocean seawater. These considerations make it difficult to assess the exact area of euxinic sediment deposition that is not affected by basinal restriction. One previous study suggested a best estimate of ~180,000 km^2 (or 0.05 percent of total seafloor (Scott et al., 2008)), but this estimate also included the most reducing portions of OMZs, which have here been included in the estimate for anoxic sediments. That area should, therefore, be considered an upper limit. Deposition rates of V in euxinic sediments are not very abundant, but a single value from the Cariaco Basin yields a flux of ~22 µmol/cm^2 × kyr (Piper and Dean, 2002). This value is similar to what is found for anoxic sediments, which is consistent with the conclusion that marine particles settling in both anoxic and euxinic environments are similarly enriched (Calvert et al., 2015; Ho et al., 2018). If the V flux is combined with the maximum euxinic seafloor area, then a maximum V flux associated with euxinic sediments of 39 Mmol/yr is obtained. However, the true value is likely substantially lower than this, given that the area of truly euxinic seafloor where the water column contains sulfide is likely much lower than what is used here.

Table 1 Authigenic V fluxes calculated for core top continental margin sediments

Sediment core top	Location	BW O_2 (μ M)	Al^ (μ g/g)	V (μ g/g)	V/Al ($\times 10^{-3}$)	Percent enrichment rel. CC*	Sed. MAR# (g/cm² kyr)	V flux (μ mol/cm² kyr)$	Ref.
TN047-20	Arabian Sea	0	17,100	43.2	2.53	112.3	10.3	4.6 ± 0.9	1
NH01 G6 C-47	Santa Barbara Basin, California Margin	1	85,900	150	1.75	46.8	65	60.9 ± 19.1	2,3
MUC29	Peru Margin	<1.5	39,300	106	2.70	94.0	28	28.2 ± 5.8	4
MUC19	Peru Margin	<1.5	22,300	227	10.2	632.3	8	30.7 ± 3.6	4
MUC39	Peru Margin	4.2	29,100	52.2	1.79	29.1	18	4.1 ± 1.8	4
TN047-22	Arabian Sea	≥5	18,100	33.3	1.84	54.6	10.3	2.4 ± 0.7	1
NH15P	Mexican Margin	5	79,000	200	2.53	112.8	12.75	26.5 ± 5.0	5
SPC	Santa Monica Basin, California Margin	5	74,400	123	1.65	38.6	18.8	12.5 ± 4.5	6
SCR-44	Santa Cruz Basin, California Margin	10	85,600	121	1.41	18.8	18.8	7.0 ± 4.4	6
TN047-30	Arabian Sea	≥10	10,200	15.2	1.49	25.3	10.3	0.6 ± 0.3	1
MUC25	Peru Margin	12.1	35,500	60.3	1.70	22.2	53	11.4 ± 6.3	4
2MC39	NW US Margin	40	54,000	104	1.93	20.8	11.25	4.0 ± 2.3	7
MUC53	Peru Margin	41.6	42,100	64.1	1.52	9.5	5.1	5.6 ± 6.4	4
WEC213	NW US Margin	50	73,700	125	1.70	6.4	12	1.8 ± 2.9	1

Table 1 (cont.)

Sediment core top	Location	BW O$_2$ (μ M)	Al$^\wedge$ (μ g/g)	V (μ g/g)	V/Al ($\times 10^{-3}$)	Percent enrichment rel. CC*	Sed. MAR$^\#$ (g/ cm^2 kyr)	V flux (μ mol/cm^2 kyr)$^\$$	Ref.
4MC33	NW US Margin	55	82,600	133	1.61	1.0	9	0.2 ± 2.3	7
BC6	San Clemente Basin, California Margin	60	83,600	103	1.22	2.9	18.8	1.1 ± 3.8	6
NH22-P	Mexican Margin	100	72,000	95.0	1.32	10.9	4.5	0.8 ± 0.8	5
8470–1	Cape Basin, Namibia	187	10,800	15.0	1.39	16.7	6.25	0.3 ± 0.2	3, 8
3BC8-1	NW African Margin	230	79,000	90.1	1.14	−4.1	2.1	-0.2 ± 0.4	1
1BC1-2	NW African Margin	240	62,000	72.0	1.16	−2.4	10.2	-0.4 ± 1.4	1
PL07- 39PC	*Cariaco Basin*	*eux*	*109,600*	*190*	*1.73*	*45.7*	*18.75*	*21.9 ± 7.0*	*9*

Sediment core from Cariaco Basin in italic to denote deposition in euxinic environment.

$^\wedge$ Al concentrations in California Margin sediments calculated assuming upper continental crust Al/Ti = 21.2 (Rudnick and Gao, 2003).

* Upper continental crust (CC) value taken from Rudnick and Gao (2003) except for NW US Margin, where sediment particles from the Columbia River are used (Morford and Emerson, 1999), and Peru Margin, where average Andean andesite is used (Scholz et al., 2011).

$^\#$ Sediment mass accumulation rates (MAR) either as reported or calculated from depositional rates using density of 2.5 g/cm^3 and porosity of 0.3 following Morford and Emerson (1999). Rates listed for TN047-22 and TN047-30 are assumed identical to TN047-20.

$^\$$ Error bars calculated assuming 10 percent total propagated uncertainty on the sedimentation rates and the Al and V concentration measurements.

References 1: (Morford and Emerson, 1999); 2: (Zheng et al., 1999); 3: (Wu et al., 2020); 4: (Scholz et al., 2011); 5: (Nameroff et al., 2004); 6: (Shaw et al., 1990); 7: (Morford et al., 2005); 8: (Riedinger et al., 2006); 9: (Piper and Dean, 2002).

2.2 Isotopic V Mass Balance

The marine mass balance of V is summarized in Figure 1, and within the uncertainties V appears to be at steady state in the modern ocean. This conclusion is also supported by the fact that V is largely conservative in the ocean and that the marine residence time of ~92 kyrs (Figure 1) is much longer than the ocean mixing time. In terms of V isotope data, two recent papers have shown that seawater and river waters are largely homogeneous with respect to V isotopes, with values of $\delta^{51}V_{SW} = +0.20 \pm 0.07‰$ (2se) and $\delta^{51}V_{RW} = -0.6 \pm 0.1‰$ (2se), respectively (Schuth et al., 2019; Wu et al., 2019a). The V isotope composition of rivers is similar to that of average upper continental crust (Wu et al., 2016), suggesting that weathering is not associated with significant net V isotope fractionation, despite riverine particles from a single river exhibiting relatively large variations (Schuth et al., 2019).

Vanadium isotope compositions of the most significant output fluxes have also recently been investigated (Wu et al., 2017; Wu et al., 2020). Here it was found that the seawater-derived (authigenic) component in oxic sediments exhibits the isotopically lightest values of $\delta^{51}V_{oxic} = 0.9 \pm 0.1‰$ (2se). This value encompasses data for both abyssal pelagic clays and oxic portions of the continental shelf, even for sediments with high total organic carbon (TOC) contents (Wu et al., 2020). Furthermore, this V isotope composition is similar to what has been found for Fe-Mn crusts (Wu et al., 2019b), which suggests that adsorption onto Fe and Mn oxyhydroxides is the primary process controlling V transfer into oxic sediments. The isotopic evidence also supports the conclusion that continental margin sediments are not associated with large V inputs to the ocean and that oxic margin sediments can most likely be considered in the same V output flux as abyssal pelagic clays.

The V isotope composition of the authigenic component in anoxic sediments is substantially heavier than in oxic sediments, with an average value for the Peru Margin and Santa Barbara Basin of $\delta^{51}V_{anoxic} = -0.5 \pm 0.1‰$ (2se). These values were interpreted to reflect that incorporation of V into settling particles in the anoxic water column was associated with an isotope fractionation factor relative to seawater of $\Delta^{51}V_{anoxic} = -0.7‰$ (Wu et al., 2020). The relatively invariant V isotope fractionation for anoxic sediments in the Peruvian OMZ was independent of sedimentation rate and V enrichment, which is consistent with the water column exhibiting only minimal V depletion relative to open ocean seawater (Ho et al., 2018). In an extension of the anoxic sediment data, the authigenic component in the euxinic Cariaco Basin sediments exhibits $\delta^{51}V_{euxinic} = -0.2 \pm 0.1‰$ (2se) and thus $\Delta^{51}V_{euxinic} = -0.4‰$ (Wu et al., 2020). In this basin the euxinic portion of the water column is depleted by

~65 percent relative to open ocean seawater (Emerson and Huested, 1991). Thus, a simple Rayleigh distillation model using the same fractionation factor during incorporation into settling particles as for anoxic sediments could account for the observed heavier V isotope compositions (Wu et al., 2020). If the assumption is that the V isotope fractionation mechanism is similar for anoxic and euxinic conditions, then the net fractionation factor associated with euxinic environments strongly depends on the degree of V depletion in the water column (Wu et al., 2020). Thus, more restricted basins might be expected to produce even smaller fractionation factors than the Cariaco Basin, with drawdown efficiency closer to 100 percent, resulting in net $\Delta^{51}V_{euxinic} = 0$‰.

The last marine V output flux is associated with adsorption onto Fe oxides in hydrothermal sediments (Trefry and Metz, 1989). To date, one conference abstract has reported V isotope values for the authigenic component in sediments both very proximal to (TAG hydrothermal field) and distal from (East Pacific Zonal Transect west of the East Pacific Rise) hydrothermal vents. Both sample sets revealed $\delta^{51}V \sim -0.3$ to 0‰ (Wu et al., 2017), suggesting either that V adsorption onto hydrothermal particles is associated with smaller isotope fractionation factors than Fe-Mn minerals in oxic sediments or that the rapid precipitation of Fe oxides from hydrothermal fluids causes almost quantitative scavenging of V from ambient seawater.

Combination of all the V isotope data for output fluxes yields an average total output flux of $\delta^{51}V \sim -0.7$‰, which is within error of the average riverine input flux (Figure 1). Furthermore, the relatively large uncertainties on some of the output fluxes allow the construction of a modern marine V isotope mass balance that is at steady state.

3 Analytical Methodologies

3.1 Chemical Separation and Isotope Measurements

Measurements of V isotope compositions are performed using multiple collector inductively coupled plasma mass spectrometry (MC-ICPMS). In order to obtain precise and accurate data, it is critical to separate V from its sample matrix, in particular Cr and Ti, because these elements both possess an isotope with mass of 50 atomic mass units that interferes with the minor ^{50}V isotope, which has an abundance of only ~0.24 percent. The separation is performed using multiple sequential ion exchange chromatography columns. Most recent studies first remove the majority of major elements using cation exchange resin (Wu et al., 2016). Subsequently, one or more anion exchange columns are used to quantitatively remove the remaining minor and trace elements, especially Cr and Ti (Nielsen et al., 2011). Studies in which V isotope compositions of natural water samples are measured also use a preconcentration column containing either a chelex resin or a so-called NOBIAS resin (Schuth et al., 2019; Wu et al., 2019a), which primarily removes most of the salt ions, thus allowing the cation and anion exchange columns to be employed subsequently.

Vanadium isotope measurements by MC-ICPMS obtain the best results when the major isotope, ^{51}V, is measured using a 10^{10} Ω resistor (Nielsen et al., 2016; Wu et al., 2016), which allows sufficient signals on ^{50}V using a conventional 10^{11} Ω resistor. Isotope compositions are measured using a standard sample bracketing procedure that corrects for instrumental drift and mass bias (Nielsen et al., 2011), although it is also possible to use admixed Fe together with standard sample bracketing to correct for instrumental mass bias (Schuth et al., 2019). Vanadium isotope compositions are reported relative to the Alfa Aesar standard solution first used at Oxford University (Nielsen et al., 2011; Prytulak et al., 2011), which has been distributed to all the labs that actively measure V isotopes in geological materials. It has been shown that different batches of high-purity V solution purchased from Alfa Aesar exhibit different V isotope compositions (Schuth et al., 2017), which prohibits generic use of Alfa Aesar V solutions as isotopic standards. However, the original Alfa Aesar solution first used at Oxford University consisted of 2 L with a concentration of 1,000 µg/ml, little of which has been consumed. Continued distribution of this solution for the foreseeable future, therefore, remains the most effective means to facilitate inter-laboratory data comparisons. Alternative V isotopic standards, such as NIST 3165, have been suggested for community use (Schuth et al., 2017;Schuth et al., 2019). However, NIST 3165 is not certified for isotopic use and has greater trace impurities than Alfa Aesar. Given that more than 10^6 µg remains of the original Oxford Alfa Aesar solution, currently available alternatives do not offer any community advantages.

The most recent studies that have adopted the methods outlined in this section report long-term external reproducibility of 0.08–0.12‰ (2sd) (Nielsen et al., 2019; Nielsen et al., 2020; Wu et al., 2016; Wu et al., 2019b; Wu et al., 2020).

3.2 Sample Treatment Procedures

For many potential sediment archives that may retain seawater V isotope variations, it is necessary to selectively dissolve only the authigenic portion of the sediment. The best means to achieve this goal depends on what type of sediment is studied. Generally, shales and other organic-rich sediments will likely have authigenic V primarily associated with the organic component. Partial dissolution experiments on United States Geological Survey shale reference materials revealed that submerging such sediments in a range of nitric acid strengths (1–5 M) and temperatures (25–130°C) did not induce notable V isotope fractionation despite variable amounts of V being released (Wu et al., 2020). It was concluded that room temperature and 3 M HNO$_3$ corresponded to the optimal partial digestion method, because this combination released the least Al relative to V (Wu et al., 2020). On the other hand, oxic sediments where V is most likely bound within Fe-Mn oxyhydroxides (Thomson et al., 1984) have been treated with a partial digestion method using 25 per cent acetic acid and 1 M hydroxylamine hydrochloride that preferentially targets these authigenic phases (Bayon et al., 2002).

4 Potential for Applications of V Isotopes to Reconstruct Past Ocean Redox

Given the isotope mass balance of V, it is likely that the V isotope composition of seawater changed in the past as a function of changes in the relative proportions of V deposited with the four different sedimentary output fluxes (Figure 1). Furthermore, it is possible to make generalized predictions regarding the potential magnitude and direction of changes in the V isotope composition of seawater under certain environmental conditions.

We can first consider an almost entirely anoxic ocean where oxic sedimentation is but a fraction of modern values. Given that V fluxes into oxic sediments are very low relative to anoxic and euxinic sediments, even 5–10 per cent oxic seafloor would likely have only a minor effect on the global V isotope budget. In addition, anoxic conditions in the deep ocean, which have been hypothesized for much of the Archean and Proterozoic periods (Canfield, 1998), would potentially have prevented rapid Fe oxide precipitation from hydrothermal fluids, thus rendering the V flux associated with such sediments relatively minor. If we assume that riverine V had the same isotope composition as today and that V isotope fractionation associated with anoxic and euxinic sediments was also the same as modern values (i.e. $\Delta^{51}V_{anoxic} = -0.7‰$ and $\Delta^{51}V_{euxinic} = -0.4‰$), then seawater would have evolved to V isotope compositions of $\delta^{51}V_{SW} = -0.3‰$ to $+0.0‰$, depending on whether euxinic or anoxic conditions, respectively, were the most prevalent. If we, furthermore, expand our considerations to include euxinic environments where V drawdown was closer to 100 percent (producing $\Delta^{51}V_{euxinic} \sim 0‰$), then seawater could even have obtained values as light as $\delta^{51}V_{SW} = -0.6‰$ in the most extreme case (Figure 3). This range of values is significantly lighter than in present day, and, in particular, predominantly euxinic conditions should result in the isotopically most different values relative to present day. If these very coarse assumptions hold true, then it might be expected that the transition away from a euxinic deep ocean that has been hypothesized around the Precambrian–Cambrian boundary

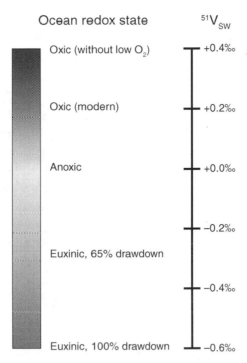

Figure 3 Schematic illustration of the V isotope composition of open ocean seawater as a function of the dominant sediment depositional environments. The most oxic case considers an ocean devoid of anoxic and euxinic environments. The lightest isotope compositions are obtained in cases where essentially all V is deposited in euxinic environments with average V drawdown of 65 percent (similar to modern Cariaco Basin) or 100 percent.

(Erwin et al., 2011) would be likely to produce substantial V isotope variation. Most likely, these variations could be detected in black shales deposited in basins that were well connected to the open ocean. However, it will be critical to find sediment sequences where V availability in the water column did not change dramatically, because the V isotope composition of anoxic/euxinic sediments likely is also controlled by degree of V uptake in the water column (e.g. Cariaco Basin). It is unclear whether only euxinic conditions are capable of causing significant V depletion within a water column, but any local changes in depositional environment between anoxic and euxinic conditions would likely induce V isotope variations that are unrelated to changes in the V isotope composition of seawater (Wu et al., 2020).

In a fully oxic ocean like the modern ocean, there would be significant V fluxes associated with oxic, anoxic, and hydrothermal sediments, whereas euxinic sediments likely would be less important. However, since fluxes associated with hydrothermal sediments most likely would be relatively constant (over million-year time scales, where ocean crust production and thus hydrothermal fluid fluxes do not change substantially), V isotope variations in seawater for geologically short-lived climate perturbations would primarily be controlled by the ratio between the anoxic and oxic output fluxes. Such events could, for example, be OAEs or mass extinction events, where the flux of V into anoxic sediments likely changed substantially (Owens et al., 2016). For example, a shutdown of marine organic carbon burial (and hence low-oxygen environments) would be associated with $\delta^{51}V_{SW}$ as high as +0.4‰, whereas predominance of anoxic over oxic V burial would result in $\delta^{51}V_{SW}$ close to 0‰ (Figure 3).

It should be stressed that the heaviest and lightest values outlined here are unlikely ever to have been produced in open ocean seawater, but they are useful endmember values that essentially define the full spectrum of possible values. Furthermore, various combinations of the different V output fluxes may produce similar V isotope values. Hence, a single V isotope composition is not necessarily diagnostic of a specific environment as suggested in Figure 3. For example, values of $\delta^{51}V_{SW} = +0.1$ could be produced by a large number of output configurations, which also include changes in the oxic output flux that are independent of ocean anoxia/euxinia. Hence, like most of the recently developed stable isotope redox proxies, V isotopes may find most utility when used in combination with other isotope proxies (e.g. Mo, U, Se, Tl, and Cr) that are sensitive to different ranges in ocean redox.

Given these considerations, the total possible variation of V isotopes in seawater over geological time is ~1.0‰, which is more than 10 times the current long-term reproducibility of V isotope measurements. Vanadium isotopes, therefore, have significant potential to place new constraints on redox changes in the oceans through Earth's history. In addition, V isotopes might also find further utility because significant V isotope variations in organic-rich shales are induced by V depletion in the water column, which is likely controlled by basinal restriction as well as global V drawdown. However, further investigations are required in order to assess the effects of basinal restriction versus global drawdown on the V isotope compositions in organic-rich sediments.

Key Papers

First high-precision V isotope studies that documented chemical separation techniques and mass spectrometric protocols:

Nielsen, S. G., Prytulak, J. and Halliday, A. N. (2011) Determination of precise and accurate 51 V / 50 V isotope ratios by MC-ICP-MS, part 1: Chemical separation of vanadium and mass spectrometric protocols. *Geostand. Geoanal. Res.*, 35, 293–306.

Prytulak, J., Nielsen, S. G. and Halliday, A. N. (2011) Determination of precise and accurate 51 V / 50 V isotope ratios by multi-collector ICP-MS, part 2: Isotopic composition of six reference materials plus the Allende chondrite and verification tests. *Geostand. Geoanal. Res.*, 35, 307–18.

Wu, F., Qi, Y. H., Yu, H. M., et al. (2016) Vanadium isotope measurement by MC-ICP-MS. *Chem. Geol.*, 421, 17–25.

Nielsen, S. G., Owens, J. D. and Horner, T. J. (2016) Analysis of high-precision vanadium isotope ratios by medium resolution MC-ICP-MS. *J. Anal. At. Spectrom.*, 31, 531–6.

Papers that documented the composition of seawater and modern marine fluxes:

Wu, F., Owens, J. D., Huang, T., et al. (2019) Vanadium isotope composition of seawater. *Geochim. Cosmochim. Acta* 244, 403–415.

Wu, F., Owens, J. D., Scholz, F., et al. (2020) Sedimentary vanadium isotope signatures in marine low oxygen bottom water conditions. *Geochim. Cosmochim. Acta*, in press.

Wu, F., Owens, J. D., Tang, L., Dong, Y. and Huang, F. (2019) Vanadium isotopic fractionation during the formation of marine ferromanganese crusts and nodules. *Geochim. Cosmochim. Acta*, 265, 371–85.

Schuth, S., Brüske, A., Hohl, S. V., et al. (2019) Vanadium and its isotope composition of river water and seawater: Analytical improvement and implications for vanadium isotope fractionation. *Chem.Geol.*, 528, 119261.

Theoretical study of V isotope fractionation:

Wu, F., Qin, T., Li, X. F., et al. (2015) First-principles investigation of vanadium isotope fractionation in solution and during adsorption. *Earth Planet. Sci. Lett.*, 426, 216–24.

References

Algeo, T. J., 2004. Can marine anoxic events draw down the trace element inventory of seawater? *Geology*, 32(12): 1057–60.

Anbar, A. D., Duan, Y., Lyons, T. W., et al., 2007. A whiff of oxygen before the Great Oxidation Event? *Science*, 317(5846): 1903–6.

Bayon, G., German, C. R., Boella, R. M., et al., 2002. An improved method for extracting marine sediment fractions and its application to Sr and Nd isotopic analysis. *Chem. Geol.*, 187: 179–99.

Beck, M., Dellwig, O., Schnetger, B. and Brumsack, H.-J., 2008. Cycling of trace metals (Mn, Fe, Mo, U, V, Cr) in deep pore waters of intertidal flat sediments. *Geochim. Cosmochim. Acta*, 72(12): 2822–40.

Bigeleisen, J. and Mayer, M. G., 1947. Calculation of equilibrium constants for isotopic exchange reactions. *J. Chem. Phys.*, 15(5): 261–7.

Brumsack, H. J. and Gieskes, J. M., 1983. Interstitial water trace-metal chemistry of laminated sediments from the Gulf of California, Mexico. *Mar. Chem.*, 14(1): 89–106.

Calvert, S. E. and Pedersen, T. F., 1993. Geochemistry of recent oxic and anoxic marine-sediments – implications for the geological record. *Mar. Geol.*, 113(1–2): 67–88.

Calvert, S. E., Piper, D. Z., Thunell, R. C. and Astor, Y., 2015. Elemental settling and burial fluxes in the Cariaco Basin. *Mar. Chem.*, 177: 607–29.

Canfield, D. E., 1998. A new model for Proterozoic ocean chemistry. *Nature*, 396(6710): 450–3.

Canfield, D. E., Zhang, S., Wang, H., et al., 2018. A Mesoproterozoic iron formation. *Proc. Natl Acad. Sci.*, 115(17): E3895–E3904.

Collier, R. W., 1984. Particulate and dissolved vanadium in the North Pacific Ocean. *Nature*, 309(5967): 441–4.

Coogan, L. A. and Dosso, S., 2012. An internally consistent, probabilistic, determination of ridge-axis hydrothermal fluxes from basalt-hosted systems. *Earth Planet. Sci. Lett.*, 323–324: 92–101.

Emerson, S. R. and Huested, S. S., 1991. Ocean anoxia and the concentrations of molybdenum and vanadium in seawater. *Mar. Chem.*, 34(3–4): 177–96.

Erwin, D. H., Laflamme, M., Tweedt, S. M., et al., 2011. The Cambrian conundrum: Early divergence and later ecological success in the early history of animals. *Science*, 334(6059): 1091–7.

Fitzsimmons, J. N., Boyle, E. A. and Jenkins, W. J., 2014. Distal transport of dissolved hydrothermal iron in the deep South Pacific Ocean. *Proc. Natl Acad. Sci.*, 111(47): 16654–61.

Fitzsimmons, J. N., John, S. G., Marsay, C. M., et al., 2017. Iron persistence in a distal hydrothermal plume supported by dissolved–particulate exchange. *Nat. Geosci.*, 10: 195.

Francois, R., 1988. A study on the regulation of the concentrations of some trace metals (Rb, Sr, Zn, Pb, Cu, V, Cr, Ni, Mn and Mo) in Saanich Inlet Sediments, British Columbia, Canada. *Mar. Geol.*, 83(1): 285–308.

Frei, R., Gaucher, C., Poulton, S. W. and Canfield, D. E., 2009. Fluctuations in Precambrian atmospheric oxygenation recorded by chromium isotopes. *Nature*, 461: 250.

Gustafsson, J. P., 2019. Vanadium geochemistry in the biogeosphere – speciation, solid-solution interactions, and ecotoxicity. *Appl. Geochem.*, 102: 1–25.

Hein, J. R., Koschinsky, A. and Halliday, A. N., 2003. Global occurrence of tellurium-rich ferromanganese crusts and a model for the enrichment of tellurium. *Geochim. Cosmochim. Acta*, 67(6): 1117–27.

Helly, J. J. and Levin, L. A., 2004. Global distribution of naturally occurring marine hypoxia on continental margins. *Deep Sea Res. 1 Oceanogr. Res. Pap.*, 51(9): 1159–68.

Ho, P., Lee, J.-M., Heller, M. I., Lam, P. J. and Shiller, A. M., 2018. The distribution of dissolved and particulate Mo and V along the U.S. GEOTRACES East Pacific Zonal Transect (GP16): The roles of oxides and biogenic particles in their distributions in the oxygen deficient zone and the hydrothermal plume. *Mar. Chem.*, 201: 242–55.

Jeandel, C., Caisso, M. and Minster, J. F., 1987. Vanadium behaviour in the global ocean and in the Mediterranean sea. *Mar. Chem.*, 21(1): 51–74.

Jenkyns, H. C., 2010. Geochemistry of oceanic anoxic events. *Geochem. Geophys. Geosyst.*, 11: Q03004, doi: 03010.01029/02009gc002788.

Kendall, B., Dahl, T. W. and Anbar, A. D., 2017. Good Golly, Why Moly? The stable isotope geochemistry of molybdenum. *Non-Traditional Stable Isotopes*, 82: 683–732.

Kumar, N., Anderson, R. F. and Biscaye, P. E., 1996. Remineralization of particulate authigenic trace metals in the middle Atlantic Bight: Implications for proxies of export production. *Geochim. Cosmochim. Acta*, 60(18): 3383–97.

Lau, K. V., Romaniello, S. J. and Zhang, F., 2019. *The Uranium Isotope Paleoredox Proxy.* Elements in Geochemical Tracers in Earth System Science. Cambridge University Press, Cambridge.

Lewan, M. D. and Maynard, J. B., 1982. Factors controlling the enrichment of vanadium and nickel in the bitumen of organic sedimentary-rocks. *Geochim. Cosmochim. Acta*, 46(12): 2547–60.

Lu, W. Y., Ridgwell, A., Thomas, E., et al., 2018. Late inception of a resiliently oxygenated upper ocean. *Science*, 361(6398): 174–7.

Lyons, T. W., Anbar, A. D., Severmann, S., Scott, C. and Gill, B. C., 2009. Tracking euxinia in the ancient ocean: A multiproxy perspective and Proterozoic case study. *Annu. Rev. Earth Planet. Sci.*, 37: 507–34.

Lyons, T. W., Reinhard, C. T. and Planavsky, N. J., 2014. The rise of oxygen in Earth's early ocean and atmosphere. *Nature*, 506(7488): 307–15.

McArthur, J. M., Algeo, T. J., van de Schootbrugge, B., Li, Q. and Howarth, R. J., 2008. Basinal restriction, black shales, Re-Os dating, and the Early Toarcian (Jurassic) oceanic anoxic event. *Paleoceanography*, 23 (4):PA4217, doi: 4210.1029/2008pa001607.

Morford, J. L. and Emerson, S., 1999. The geochemistry of redox sensitive trace metals in sediments. *Geochim. Cosmochim. Acta*, 63(11–12): 1735–50.

Morford, J. L., Emerson, S. R., Breckel, E. J. and Kim, S. H., 2005. Diagenesis of oxyanions (V, U, Re, and Mo) in pore waters and sediments from a continental margin. *Geochim. Cosmochim. Acta*, 69(21): 5021–32.

Nameroff, T. J., Balistrieri, L. S. and Murray, J. W., 2002. Suboxic trace metal geochemistry in the eastern tropical North Pacific. *Geochim. Cosmochim. Acta*, 66(7): 1139–58.

Nameroff, T. J., Calvert, S. E. and Murray, J. W., 2004. Glacial-interglacial variability in the eastern tropical North Pacific oxygen minimum zone recorded by redox-sensitive trace metals. *Paleoceanography*, 19(1).

Nance, W. B. and Taylor, S. R., 1976. Rare earth element patterns and crustal evolution – I. Australian post-Archean sedimentary rocks. *Geochim. Cosmochim. Acta*, 40(12): 1539–51.

Nielsen, S. G., Auro, M., Righter, K., et al., 2019. Nucleosynthetic vanadium isotope heterogeneity of the early solar system recorded in chondritic meteorites. *Earth Planet. Sci. Lett.*, 505: 131–40.

Nielsen, S. G., Bekaert, D. V. and Auro, M., 2020. Isotopic evidence for the formation of the Moon in a giant impact. *Nature*, in review.

Nielsen, S. G., Owens, J. D. and Horner, T. J., 2016. Analysis of high-precision vanadium isotope ratios by medium resolution MC-ICP-MS. *J. Anal. At. Spectrom.*, 31: 531–6.

Nielsen, S. G., Prytulak, J. and Halliday, A. N., 2011. Determination of precise and accurate 51 V / 50 V isotope ratios by MC-ICP-MS, part 1: Chemical separation of vanadium and mass spectrometric protocols. *Geostand. Geoanal. Res.*, 35(3): 293–306.

Nielsen, S. G., Rehkämper, M. and Prytulak, J., 2017. Investigation and application of thallium isotope fractionation. *Rev. Mineral. Geochem.*, 82(1): 759–98.

Ohnemus, D. C., Rauschenberg, S., Cutter, G. A., et al., 2017. Elevated trace metal content of prokaryotic communities associated with marine oxygen deficient zones. *Limnol. Oceanogr.*, 62(1): 3–25.

Ostrander, C. M., Nielsen, S. G., Owens, J. D., et al., 2019a. Fully oxygenated water columns over continental shelves before the Great Oxidation Event. *Nat. Geosci.*, 12(3): 186–91.

Ostrander, C. M., Sahoo, S. K., Kendall, B., et al., 2019b. Multiple negative molybdenum isotope excursions in the Doushantuo Formation (South China) fingerprint complex redox-related processes in the Ediacaran Nanhua Basin. *Geochim. Cosmochim. Acta*, 261: 191–209.

Owens, J. D., 2019. *Application of Thallium Isotopes: Tracking Marine Oxygenation through Manganese Oxide Burial*. Elements in Geochemical Tracers in Earth System Science. Cambridge University Press, Cambridge.

Owens, J. D., Lyons, T. W., Li, X., et al., 2012. Iron isotope and trace metal records of iron cycling in the proto-North Atlantic during the Cenomanian-Turonian oceanic anoxic event (OAE-2). *Paleoceanography*, 27(3).

Owens, J. D., Reinhard, C. T., Rohrssen, M., Love, G. D. and Lyons, T. W., 2016. Empirical links between trace metal cycling and marine microbial ecology during a large perturbation to Earth's carbon cycle. *Earth Planet. Sci. Lett.*, 449: 407–17.

Piper, D. Z. and Dean, W. E., 2002. Trace-element deposition in the Cariaco Basin, Venezuela Shelf, under sulfate-reducing conditions – a history of the local hydrography and global climate, 20 ka to the present. US Geological Survey: Professional paper 1670.

Prytulak, J., Nielsen, S. G. and Halliday, A. N., 2011. Determination of precise and accurate 51 V / 50 V isotope ratios by multi-collector ICP-MS, part 2: Isotopic composition of six reference materials plus the Allende chondrite and verification tests. *Geostand. Geoanal. Res.*, 35(3): 307–18.

Rehkämper, M. and Nielsen, S. G., 2004. The mass balance of dissolved thallium in the oceans. *Mar. Chem.*, 85: 125–39.

Riedinger, N., Kasten, S., Gröger, J., Franke, C. and Pfeifer, K., 2006. Active and buried authigenic barite fronts in sediments from the Eastern Cape Basin. *Earth Planet. Sci. Lett.*, 241(3): 876–87.

Rudnick, R. L. and Gao, S., 2003. Composition of the Continental Crust. In: H. D. Holland and K. K. Turekian (eds), *Treatise on Geochemistry*. Pergamon, Oxford, pp. 1–64.

Rudnicki, M. D. and Elderfield, H., 1993. A chemical model of the buoyant and neutrally buoyant plume above the TAG vent field, 26 degrees N, Mid-Atlantic Ridge. *Geochim. Cosmochim. Acta*, 57(13): 2939–57.

Rue, E. L., Smith, G. J., Cutter, G. A. and Bruland, K. W., 1997. The response of trace element redox couples to suboxic conditions in the water column. *Deep Sea Res. 1 Oceanogr. Res. Pap.*, 44(1): 113–34.

Scholz, F., Hensen, C., Noffke, A., et al., 2011. Early diagenesis of redox-sensitive trace metals in the Peru upwelling area – response to ENSO-related oxygen fluctuations in the water column. *Geochim. Cosmochim. Acta*, 75 (22): 7257–76.

Schuth, S., Brüske, A., Hohl, S. V., et al., 2019. Vanadium and its isotope composition of river water and seawater: Analytical improvement and implications for vanadium isotope fractionation. *Chem. Geol.*, 528: 119261.

Schuth, S., Horn, I., Bruske, A., Wolff, P. E. and Weyer, S., 2017. First vanadium isotope analyses of V-rich minerals by femtosecond laser ablation and solution-nebulization MC-ICP-MS. *Ore Geol. Rev.*, 81: 1271–86.

Scott, C., Lyons, T. W., Bekker, A., et al., 2008. Tracing the stepwise oxygenation of the Proterozoic ocean. *Nature*, 452(7186): 456–U455.

Shaw, T. J., Gieskes, J. M. and Jahnke, R. A., 1990. Early diagenesis in differing depositional environments: The response of transition metals in pore water. *Geochim. Cosmochim. Acta*, 54(5): 1233–46.

Shiller, A. M. and Boyle, E. A., 1987. Variability of dissolved trace metals in the Mississippi river. *Geochim. Cosmochim. Acta*, 51(12): 3273–7.

Shiller, A. M. and Mao, L., 2000. Dissolved vanadium in rivers: Effects of silicate weathering. *Chem. Geol.*, 165(1): 13–22.

Southam, J. R. and Hay, W. W., 1977. Time scales and dynamic models of deep-sea sedimentation. *J. Geophys. Res.*, 82(27): 3825–42.

Stüeken, E. E., Buick, R. and Anbar, A. D., 2015. Selenium isotopes support free O2 in the latest Archean. *Geology*, 43(3): 259–62.

Tagliabue, A., Bopp, L., Dutay, J.-C., et al., 2010. Hydrothermal contribution to the oceanic dissolved iron inventory. *Nat. Geosci.*, 3(4): 252–6.

Teagle, D. A. H., Bickle, M. J. and Alt, J. C., 2003. Recharge flux to ocean-ridge black smoker systems: A geochemical estimate from ODP Hole 504B. *Earth Planet. Sci. Lett.*, 210: 81–9.

Thomson, J., Carpenter, M. S. N., Colley, S., et al., 1984. Metal accumulation rates in Northwest Atlantic pelagic sediments. *Geochim. Cosmochim. Acta*, 48(10): 1935–48.

Trefry, J. H. and Metz, S., 1989. Role of hydrothermal precipitates in the geochemical cycling of vanadium. *Nature*, 342(6249): 531–3.

Tribovillard, N., Algeo, T. J., Lyons, T. and Riboulleau, A., 2006. Trace metals as paleoredox and paleoproductivity proxies: An update. *Chem. Geol.*, 232 (1–2):12–32.

Wanty, R. B. and Goldhaber, M. B., 1992. Thermodynamics and kinetics of reactions involving vanadium in natural systems: Accumulation of vanadium in sedimentary rocks. *Geochim. Cosmochim. Acta*, 56(4): 1471–83.

Wehrli, B. and Stumm, W., 1989. Vanadyl in natural-waters – adsorption and hydrolysis promote oxygenation. *Geochim. Cosmochim. Acta*, 53(1): 69–77.

Wu, F., Owens, J. D., Huang, T., et al., 2019a. Vanadium isotope composition of seawater. *Geochim. Cosmochim. Acta*, 244: 403–15.

Wu, F., Owens, J. D., Nielsen, S. G., German, C. R. and Mills, R. A., 2017. V isotope composition in modern marine hydrothermal sediments, AGU Fall Meeting. AGU, New Orleans, pp. V11A–0330.

Wu, F., Owens, J. D., Scholz, F., et al., 2020. Sedimentary vanadium isotope signatures in marine low oxygen bottom water conditions. *Geochim. Cosmochim. Acta*, in press.

Wu, F., Owens, J. D., Tang, L., Dong, Y. and Huang, F., 2019b. Vanadium isotopic fractionation during the formation of marine ferromanganese crusts and nodules. *Geochim. Cosmochim. Acta*, 265: 371–85.

Wu, F., Qi, Y. H., Yu, H. M., et al., 2016. Vanadium isotope measurement by MC-ICP-MS. *Chem. Geol.*, 421: 17–25.

Zheng, Y., Anderson, R. F., van Geen, A. and Kuwabara, J., 2000. Authigenic molybdenum formation in marine sediments: A link to pore water sulfide in the Santa Barbara Basin. *Geochim. Cosmochim. Acta*, 64(24): 4165–78.

Acknowledgements

This manuscript benefited from insightful reviews by Julie Prytulak and Jeremy Owens as well as editorial handling by Tim Lyons. SGN acknowledges funding from NSF (OCE 1829406) and NASA (NNX16AJ60 G).

Cambridge Elements ☰

Elements in Geochemical Tracers in Earth System Science

Timothy Lyons
University of California

Timothy Lyons is a Distinguished Professor of Biogeochemistry in the Department of Earth Sciences at the University of California, Riverside. He is an expert in the use of geochemical tracers for applications in astrobiology, geobiology and Earth history. Professor Lyons leads the 'Alternative Earths' team of the NASA Astrobiology Institute and the Alternative Earths Astrobiology Center at UC Riverside.

Alexandra Turchyn
University of Cambridge

Alexandra Turchyn is a University Reader in Biogeochemistry in the Department of Earth Sciences at the University of Cambridge. Her primary research interests are in isotope geochemistry and the application of geochemistry to interrogate modern and past environments.

Chris Reinhard
Georgia Institute of Technology

Chris Reinhard is an Assistant Professor in the Department of Earth and Atmospheric Sciences at the Georgia Institute of Technology. His research focuses on biogeochemistry and paleoclimatology, and he is an Institutional PI on the 'Alternative Earths' team of the NASA Astrobiology Institute.

About the Series

This innovative series provides authoritative, concise overviews of the many novel isotope and elemental systems that can be used as 'proxies' or 'geochemical tracers' to reconstruct past environments over thousands to millions to billions of years—from the evolving chemistry of the atmosphere and oceans to their cause-and-effect relationships with life.

Covering a wide variety of geochemical tracers, the series reviews each method in terms of the geochemical underpinnings, the promises and pitfalls, and the 'state-of-the-art' and future prospects, providing a dynamic reference resource for graduate students, researchers and scientists in geochemistry, astrobiology, paleontology, paleoceanography and paleoclimatology.

The short, timely, broadly accessible papers provide much-needed primers for a wide audience—highlighting the cutting-edge of both new and established proxies as applied to diverse questions about Earth system evolution over wide-ranging time scales.

Cambridge Elements ☰

Elements in Geochemical Tracers in Earth System Science

Elements in the Series

The Uranium Isotope Paleoredox Proxy
Kimberly V. Lau, Stephen J. Romaniello and Feifei Zhang

Triple Oxygen Isotopes
Huiming Bao

Application of Thallium Isotopes
Jeremy D. Owens

Earth History of Oxygen and the iprOxy
Zunli Lu, Wanyi Lu, Rosalind E. M. Rickaby and Ellen Thomas

Selenium Isotope Paleobiogeochemistry
Eva E. Stüeken and Michael A. Kipp

The TEX$_{86}$ Paleotemperature Proxy
Gordon N. Inglis and Jessica E. Tierney

The Pyrite Trace Element Paleo-Ocean Chemistry Proxy
Daniel D. Gregory

Calcium Isotopes
Elizabeth M. Griffith and Matthew S. Fantle

Pelagic Barite: Tracer of Ocean Productivity and a Recorder of Isotopic Compositions of Seawater S, O, Sr, Ca and Ba
Weiqi Yao, Elizabeth Griffith and Adina Paytan

Vanadium Isotopes: A Proxy for Ocean Oxygen Variations
Sune G. Nielsen

A full series listing is available at: www.cambridge.org/EESS

Printed in the United States
By Bookmasters